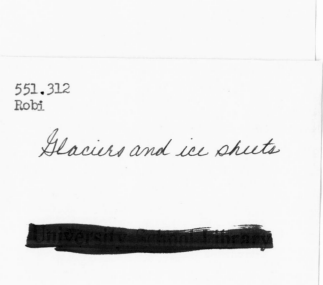

PLANET EARTH

GLACIERS AND ICE SHEETS

Gordon de Q. Robin

The Bookwright Press
New York · 1984

PLANET EARTH

First published in the United States in 1984 by
The Bookwright Press
387 Park Avenue South
New York, N.Y. 10016

First published in Great Britain in 1984 by
Wayland (Publishers) Ltd
49 Lansdowne Place, Hove
East Sussex BN3 1HF, England

ISBN 0-531-03801-7
Library of Congress Catalogue Card Number
84-70748

Printed in Italy by
G. Canale & C.S.p.A., Turin

Contents

4

Mountain glaciers

Ice in a refrigerator is hard and cold. Unless the cold parts of the refrigerator are defrosted from time to time, large lumps of ice start to fill the refrigerator. The same thing takes place in nature in many cold countries, where vast masses of ice build up to form **glaciers.** They can be seen from far away as beautiful, steep, white rivers of ice, flowing down from high mountain peaks.

In many countries winter snow only stays on the ground for a few days. In colder countries it can stay for some months before the summer heat comes to melt it away. If the snow is very deep, perhaps because it has been blown into a deep valley on the side of a mountain, the summer heat may not be enough to melt all the snow away, even though the daytime temperatures are quite warm.

Sometimes, a snowdrift that did not melt during a cold summer will melt away the following year if the summer is warmer. However, if the summers become steadily colder over many years, that snowdrift will become thicker and thicker, until it starts to flow down the mountain as a glacier. For this to happen, the ice must build up to a thickness of at least 20 meters (66 feet), that is, the height of a five-story building. Most mountain glaciers are much thicker than that—often 200 to 400 meters thick (650 to 1,300 feet) —so they could cover a large skyscraper 50 to 100 stories high.

Melting ice

Lower down the mountain, the glaciers begin to melt, and at the end of glaciers, a cold stream of water flows from the melting ice. These grey-green streams flow strongly in summer.

When you climb a mountain or go up in a plane in clear weather, you notice that the air becomes colder. For every 100 meters higher you go, the air is

Left *Glaciers flowing down the steep face of the Bernina, in the Swiss Alps.*

around 1°C colder (or 5.4°F per 1,000 feet). So if the air temperature at the top of a glacier is just below freezing, and the end of the glacier is 500 meters lower down (1,640 feet), the air above the end of the glacier will be several degrees above freezing. This is warm enough to melt quite a lot of ice during the day, especially if a strong wind blows lots of warmer air over the glacier. This could

Scientists measure the summer ice melt on a glacier.

The melting ice of this glacier forms a spectacular waterfall.

melt three centimeters (one and a quarter inches) of ice or more in a day, which is equal to one good winter snowfall.

If you have to be out in the hot summer sun, you will feel cooler if you wear white clothes than you would in black, because dark colors absorb more of the sun's heat. White reflects the sun's heat away. It is the same on a glacier. This is because fresh white snow absorbs only one fifth to one tenth of the sun's heat—the rest is reflected upwards. However, if the snow surface is dirty with dust, or if it is wet, it can absorb one half or more of the sun's heat. Ice also absorbs over one half of the heat from the sun.

The effect of the seasons

At the end of the winter, when snow covers the mountainsides and glaciers, most of the sun's heat is reflected back into the sky. This keeps the air temperature cold, even though it is often sunnier than in the autumn. This gives lovely weather for snow sports when the sun is out.

As spring advances, warmer winds and more sunshine will finally start to melt the surface of the snow. This absorbs more heat than dry snow, and once it starts to melt, the snow disappears quickly.

As melting continues, the snow becomes soggy and waterlogged, and water runs away in streams. On mountain glaciers, this happens first on the lower part of the glacier, and uncovers the ice underneath. During summer, the ice surface extends higher up the glacier to a line, above which the snow remains. At the end of the summer this **snow line** divides the lower part of the glacier from the upper. Above the snow line, even though some snow may have melted, some of the snow from the previous winter is left behind. The glacier flow slowly carries this snow and ice down to the lower part of the glacier, where it melts.

Melting and refreezing of snow

As winter temperatures over glaciers are well below the freezing point of water (0°C or 32°F), the upper layers of the glacier are chilled below freezing to a

High up on a glacier on Mt. Blanc, France, the snow line can be seen clearly.

A stream of meltwater emerges from a tunnel under the glacier.

depth of around 10 meters (33 feet). In summer, as water from melting surface snow trickles down into the colder snow, it warms up until the whole glacier is again at the melting point. Such glaciers are called **temperate glaciers**, because they are found in mountains in the more temperate climates of the world. These are the glaciers that have done most to sculpt out the valleys, cirques and fjords that we see in temperate countries.

The refreezing of **meltwater** in snow, and evaporation from small crystals of ice that condenses on large crystals, soon turn loose snow into a solid mass of crystals that stick together. This is known as **firn** and although it is solid, air and water can still flow between crystals. Deeper down, the firn changes to solid ice, through which air and water cannot move freely. The water then collects into larger streams, which melt tunnels in the glacier, carrying the water to the bottom of the glacier. There the water travels in channels between the ice and rock, to emerge as the stream at the end of the glacier.

The Athabasca Glacier in Alberta, Canada, has sculpted a deep valley between the mountain peaks.

The movement of glaciers

It is less than 200 years since people started to measure how fast temperate glaciers move, although farmers in Europe and Iceland must have known before then that they do move, because the ends of glaciers would encroach on their farmland when it had been unusually cold over many years.

The flow of ice

If you wanted to see how a glacier was moving, one way would be to lay a row of marker stones across the glacier, between two marks on the sides of the glacier. At the end of a year, if you put another row of stones across the glacier between the same marks on the sides, by measuring the distance between the first and second line of stones, you could see how far the ice had moved in that time. On a large glacier, perhaps one kilometer wide (3,280 feet), and 10 or 20 kilometers long (6 to 12 miles), the rocks in the middle of the glacier might have moved 100 meters (328 feet) or more. The rocks across most of

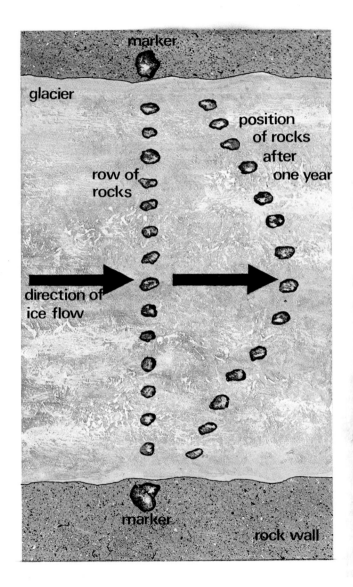

the glacier would move at this speed, but at the very edge they might only have moved half as much. If you could watch a medium-sized glacier day by day, you would see the edge of the ice move slowly past the rock wall, perhaps by ten centimeters a day (four inches).

To find out what happens inside glaciers, engineers have drilled deep holes right through the ice to the **bedrock.** With special instruments they measure how the ice moves—in much the same way as with stones on the surface. In the upper part of the drill hole they find that the ice moves almost as fast as on the surface. It is only in the lower part that the ice slows up, and at the bottom the ice slides slowly over the rock, as it did past the walls on the surface.

How glaciers move

If you take a cold ice-cube out of the freezer, it may stick and freeze to your fingers, because it is so cold. After a couple of minutes at room temperature though, the surface of the ice-cube will become wet, and if you push it along a table it will slide easily on this thin film of water.

Under a temperate glacier, a thin film of water between the ice and rock also helps it to slide, but unlike the table, the rock is never completely smooth. Some parts of the ice slide easily, pushing other sections of the ice against bumps on the bedrock with great pressure. This extra

There is often a dangerous gap between a glacier's side and the rock wall.

13

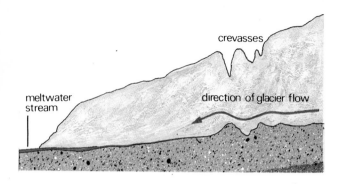

crevasses

meltwater stream

direction of glacier flow

pressure and the rubbing of ice against rock helps to melt the ice and to keep it sliding. The very strong pressure also bends the ice so that it flows around and over any bumps in the bedrock.

Bending of the ice as it flows over bumps can make cracks appear on the top of the glacier, even when the ice is hundred of meters deep. These cracks, or **crevasses,** can be very large, up to 30 meters deep (100 feet), and perhaps 6 meters wide (20 feet). During heavy snowfalls, soft snow can build snowdrifts across these crevasses, and cover them completely. However, the snow bridge may not be strong enough to take the weight of a person, so only people with experience, who know where snow-hidden crevasses may be present, can lead others safely across glaciers.

Another effect of the pressure of ice on rocks can be seen with a simple experiment at home. If you put a melting ice-cube on a small platform and hang a wire across it with heavy weights on either end, you will see the wire move slowly down through the ice. If you try to pull the wire out, you will find you cannot, because the space behind the wire has been refilled by ice. The pressure of the wire has caused the ice to melt, and the water has moved around and refrozen behind the wire. The same thing happens under glaciers where the ice is at melting point. Some of the ice being forced against a bump melts, and the water flows around the bump, where it freezes again on the other side. Both the bending around the bumps, and the melting and refreezing water helps the ice to slide over uneven rock-beds. This sliding is faster in spring and summer, when melting water from the glacier surface forces its way down to the bedrock and the water pressure tries to lift up the ice. This is why mountain glaciers slide much more in summer than in winter.

Heavy snowfalls can create snow bridges across a crevasse.

Mountain erosion

Have you ever wondered why mountain-sides are so steep that some of the rock walls rise straight up? This is because rocks break off the mountainside and fall into the valley below. After a very long time, the pile of rocks against the bottom of a mountain will have grown to form a steep pile of rocks, called **scree,** which buries the lower part of the mountain. However, when a glacier flows down a mountain valley, any rocks that fall on to

it are carried along by the slowly moving ice. Most of these rocks stay close to the sides of the glacier, where they fell. Above the snow line, you do not see many rocks, because they are soon buried beneath the snow. However, as the ice below the snow line melts away year after year, more and more of the rocks in the ice come into view. These long strips of rocks are called **moraines.** If two glaciers join to form a larger glacier, moraines from the sides of the smaller glaciers join to form a central or **medial moraine** in the larger glacier. Sometimes several glaciers feed into one large glacier and form several parallel medial moraines.

Frost shattering

Rocks that fall on the glacier have usually broken off the mountainside because of cracks opened up by frost. When water is frozen in cracks in the rock, it expands and forces the cracks open, in the same way that freezing water can burst pipes in a house. When

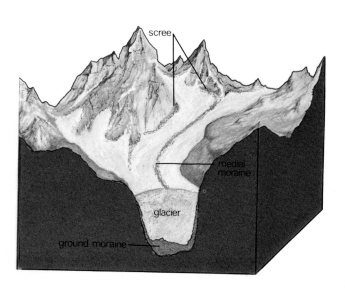

scree

medial moraine

glacier

ground moraine

Several parallel moraines are carried along in the ice at Glacier Bay, Alaska.

Glacier-smoothed rock below the Franz Josef Glacier, New Zealand.

fallen rocks get buried in the moving ice at the edge of glaciers, they scratch and grind away at the rock walls as they move along. This happens because ice at the bottom and sides of temperate glaciers is melting. As the ice around the rocks melts away, they are pressed against the sides and bottom of the glacier. The melting is caused by a combination of heat from the Earth itself, and by friction of the ice and rock rubbing together.

It is sometimes possible to crawl along a tunnel beneath a glacier, where it flows over a step in the bedrock. People in such tunnels have seen and measured the amount of grinding by rocks on the glacier bed. Fortunately, there are many places where glaciers have retreated, and you can see the bedrock without going under a glacier. The bedrock is marked with scratches and grooves of all sizes, marks from big rocks that rolled over against the bedrock, and the polishing effect where the ice rubbed fine grains of rock dust against the bedrock. Glaciers can also pluck rocks out of the bedrock, providing a further supply to scratch and grind out the glacier valleys.

Frost-shattered rock litters the surface of the Lolofond Glacier, Kashmir.

Rock flour

Of course, as rocks in the ice grind away at the bedrock and glacier walls, they wear themselves away, and break up into smaller pieces. These smaller pieces break up again and again, until they are fine enough to be carried away by the thin film of water between the ice and bedrock. These particles of rock are almost too fine to be seen by the naked eye, but there are so many of them that they are often called **rock flour.** It is the rock flour in the stream coming out from

The foot of this glacier is almost buried in rocks, and the meltwater is thick with rock flour.

beneath a glacier that gives it a color, usually grey-green or pale brown.

When you see a fast-running stream coming out from beneath a glacier, carrying all this rock flour, you can realize that a great deal of rock grinding takes place beneath a glacier. By measuring the amount of rock flour that comes out, we can work out how much the glacier valley is being eroded. Under some glaciers it can be as much as one meter (3.28 feet) in a thousand years. This may not seem like much, but since many of the glaciers we see today may have been active for more than one million years, they have had a long time to carve out deep U-shaped valleys and fjords, and create a great deal of magnificent scenery.

Deposited rocks

Although a lot of rocks carried along by glaciers are ground up into rock flour, other rocks are not destroyed, either because they are very hard or because they fell too far from the edge of the glacier to be ground against the bedrock. When they reach the end of the glacier, and if the glacier has not changed in

length for some time, such rocks build up into a big pile across the end that we call an **end moraine.** Some glaciers have several end moraines that mark the positions that the glacier has reached at different times in the past. In areas where a great deal of rock falls on a glacier's surface, such as in the Himalayas, the lowest part of the glacier can be entirely covered by rocks as the

The amount of rock moved by a glacier can be seen when the ice retreats.

ice has melted away beneath them. These layers may be from less than one meter to several meters in thickness (from one to several yards). In these cases it can be very difficult to tell where the glacier ends beneath the rock layer.

A great heap of end moraine marks the position that a glacier reached in the past.

Icecaps and ice sheets

If mountain glaciers do not melt away as they reach the foot of the mountain, they flow out over the surrounding country to form **piedmont glaciers.** *(Piedmont* is French for "foot of the mountain.") Although the slope of piedmont glaciers is much less than that of scree slopes around a mountain, if they spread out a long way the ice close to the mountain becomes quite thick.

A piedmont glacier flows over the foothills.

The mountains of N.E. Greenland are nearly buried under the ice sheet.

If the ice has become so thick that the mountain is almost buried, we call it an **icecap.** This term is used to describe all more or less circular glaciers from around 10 to 200 kilometers across (6 to 124 miles). There are five icecaps on Iceland which are larger than 200 square kilometers (77 square miles).

Ice sheets, which are still larger than ice- caps, cover most of Greenland and Antarctica. The Antarctic ice sheet is the biggest in the world, and is larger in area than either the United States or Australia.

The size to which an icecap or ice sheet grows depends much more on where the glacier ice melts than on how heavy is the snowfall. This is because without melting, the ice will keep growing. In the Antarctic, the annual snowfall on the high central region is less than the annual rainfall on many deserts.

An icecap on a plateau will flow off the edge of the plateau to melt at lower levels, like a mountain glacier. If the climate is cold, an island icecap may grow down to the shoreline. The sea water then melts the base of the ice flowing off the land. The ice above then breaks off to form the vertical ice cliffs that are often found around many small Arctic icecaps.

An Antarctic glacier flows out on to the sea.

Ice shelves and icebergs

Because the Antarctic continent lies around the South Pole, little of the ice sheet melts before it flows off the land. Even then the surrounding oceans are so cold that generally the ice is not melted as it flows across the shoreline. It continues to move out to a depth where it can float on the sea as a thick

An Antarctic piedmont glacier. The ice is forming cliffs where the glacier flows into the sea.

slab of ice, called an **ice shelf,** which is attached to the inland ice sheet. Ice shelves are mostly from 200 to 400 meters thick (650 to 1,300 feet), but can be over 1,000 meters (3,280 feet) thick in places.

The huge Ross ice shelf of Antarctica towers over the sea.

Icebergs are formed when large masses of ice break away from glaciers or ice shelves. Those that break off from ice shelves are large and flat, and are called **tabular icebergs.** The largest tabular iceberg found was 150 by 50 kilometers across (93 by 31 miles), and around 300 meters thick (1,000 feet). This is larger than the county of Kent in England, and over twice the size of the State of Rhode Island. If this iceberg were melted, it would provide enough water to supply all of London for 800 years. Indeed, for several years now the possibility of towing icebergs to some of the desert areas of the world, such as Western Australia, Peru or Arabia, has been discussed by engineers and scientists.

Measuring the ice

Because the ice sheets that cover Antarctica and Greenland are much larger than icecaps, they are also very thick. Ice thickness is measured by echo sounding, in much the same way that ocean depths are measured. One method is to set off an explosion in the ice and then time how long it takes for an echo to come back from the bedrock. Echoes have also been obtained using a radar system on an aircraft flying over the ice sheet.

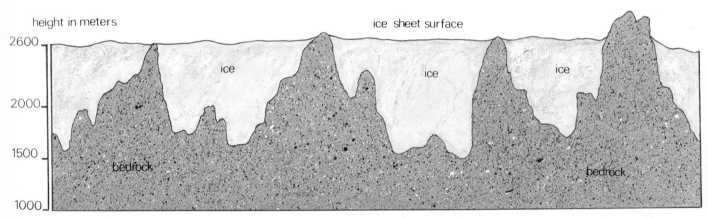

height in meters

ice sheet surface

2600

ice   ice   ice

2000

1500

bedrock   bedrock

1000

A cross-section through part of the Transantarctic Mountains.

Antarctic icebergs reflecting the afternoon sun.

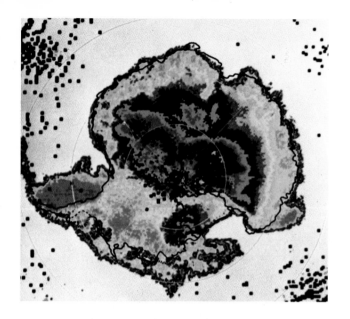

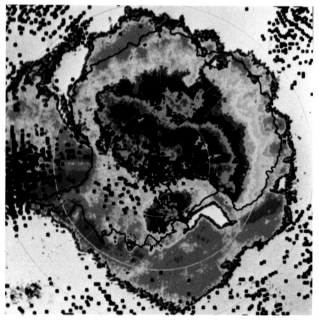

The temperature of ice sheets

The Antarctic ice sheet is so high that the surface snow or firn never warms up to melting point. The average temperature at the South Pole (2,800 meters or 9,186 feet above sea level) is close to -50°C (-58°F). In summer the monthly average rises to -28°C (-18.4°F), and in winter the monthly average falls to -64°C (-83.2°F). The temperature at the bottom of thick polar ice sheets is much warmer than at the surface. In some areas it is at the melting point of ice, while elsewhere it may still be below freezing.

The movement of ice sheets

Cold ice is much stiffer than warmer ice and it does not slide over bedrock. Also, snow piles up on ice sheets so that they

Left *These satellite images of Antarctica show how much the ice increases after the Antarctic summer* (top). *The blue areas inland are thick glaciers; yellow and red represent less thick ice.*

are highest at the center and slope out-wards towards the coasts. This makes them move like big thick sheets of cold ice being dragged outwards by gravity in the direction of their surface slope, towards the coastline. Dragging across hills and valleys in the bedrock causes bumps to appear on the surface.

Inland, on large ice sheets, most of the ice movement takes place within the ice, rather than by the ice sliding over the rocky floor. In thinner ice near the coast, valleys in the rock floor make the ice flow as a glacier rather than as a sheet.

Between these coastal glaciers, mountain ranges may block the ice flow. If the bedrock is too low for mountains to show above the ice, drainage channels carry fast-moving glaciers called **ice streams,** either side of which thinner ice moves slowly as a sheet. Many of these glaciers and ice streams may move at several hundred meters (or yards) each year, by sliding over the bedrock, on the water between the ice and rock.

The rounded surface of an ice sheet as it passes over a hill.

Ice ages

If the climate becomes colder, many mountain glaciers will extend beyond the mountains to form piedmont glaciers. If these keep on growing, icecaps are formed, and if it gets still colder, many of the icecaps will join up into a vast continental ice sheet. This all takes a very long time to happen—perhaps ten or twenty thousand years. That is around three hundred times as long as you are likely to live.

When the Earth was colder, some twenty thousand years ago, many small icecaps in the Arctic had grown so large that they joined into vast ice sheets that carried the ice far to the south, into warmer climates closer to the equator. In North America the ice sheet spread from the Arctic Islands to well south of the Great Lakes. In Europe the ice sheet extended from northern Norway and Spitsbergen to the English Midlands, Ireland, and just north of the Alps in central Europe. There was so much snow and ice over the Earth that it reflected a lot more of the sun's heat back into space, and the whole climate of the Earth became cooler. Even at the equator, the temperature over many lands was around 5°C (9°F) colder than today, but this is still quite warm for someone used to living in London or New York. These cold periods in the Earth's history are known as **ice ages.** There have been around ten ice ages during the last million or so years, varying in length from around 40,000 years to over 100,000 years. In between ice ages, or glacial periods, there are warmer

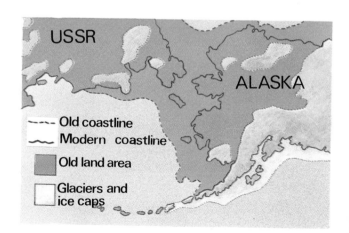

USSR

ALASKA

----- Old coastline
— Modern coastline
Old land area
Glaciers and ice caps

periods called **interglacial periods.** We are at present in an interglacial period that has so far lasted almost 10,000 years. Before that the Earth was recovering from an ice age that lasted around 50,000 years or more.

Material moved by ice sheets

Ice sheets do not erode bedrock as fast as mountain glaciers, because there tend to be no mountains above the ice sheet to supply frost-shattered rocks to grind away at bedrock. However, when ice sheets grew towards the equator during the ice ages, they moved over vast areas of land covered by soil and loose rocks. The ice picked up much of this material and moved it, sometimes over great distances.

Some rare types of rock are found only in one place, but when loose rocks of the same material are found hundreds of miles away, something must have carried them to the new site. By far the best explanation for this and for the huge amounts of soil, clay and rock deposited in central Europe and North America is that this **glacial till** as it is called, has been carried there by the ice.

Ice sheets can carry huge amounts of material over considerable distances.

The huge plains that cover much of East and West Germany and Poland, and extend into Russia, and places in America south of the Great Lakes, consist of glacial till tens or hundreds of meters (yards) thick—deposited there by the ice ages of the past million years or more.

Farther north in Canada and Europe we find very little soil or loose rock over vast areas. Hard rocks like granite have been cleared of loose soil and rock by the advancing ice sheets of the ice ages. The thin soil and vegetation there has developed over the past 10,000 years since the last retreat of the continental ice sheets. In these regions, hollows scooped out by ice action in softer rocks are now filled by lakes, and the time has not been long enough to fill these lakes with mud and clay from the surrounding land surface. Scratch marks and polishing by the rocks and rock particles in former ice sheets are common on bedrock in this region.

The surface of the rock in this valley has been smoothed and scored by a former glacier.

Nearer to the equator, where the ice sheets have deposited vast amounts of material, the landscape is quite different. Moraine deposits at the end and sides of the ice sheets may form a series of hills a few tens of meters (yards) high, undulating across the countryside. Some special forms of deposits laid down beneath ice sheets are also found. One example is the **drumlin** that has a special streamlined form caused by the ice flow as the material was deposited.

The history of ice ages

Until the early part of this century, our knowledge of former ice ages came from studies of the rocks deposited on land. The evidence showed that there had been at least four ice ages in the last two million years. But obviously as the ice sheet of each ice age advances, it may pick up moraines and till from an earlier ice age, and deposit them somewhere else.

Unlike this moraine and till, the fall of mud to the sea floor was not interrupted by ice ages. This marine mud was formed by dust carried by wind and rivers from the land to the sea. It also contains the remains of plants and animals that lived in the sea. This material falls slowly to the sea floor, perhaps taking several hundreds of years to form a layer one centimeter (less than half an inch) thick. By studying mud cores from drilling in the sea bed, we find that there have been two or three times as many ice ages as were estimated by studying glacial deposits on land. We have learned a lot more about ice ages by studying these muds than we discovered by studying moraines and glacial till. At still earlier times in the Earth's history there were other glacial periods on other continents, but less is known about them than of the ice ages of the last two million years.

Although marine muds are very useful for studying the ice ages themselves, it is only by looking at the deposits of glacial till on the land itself that we can learn how much of the continents was covered with ice during the ice ages. In all, at the maximum extent of the ice ages, the ice sheets covered an area around three times greater than the present-day ice sheets of Antarctica and Greenland.

Ice sheets, sea level and debris

When a kettle boils, the invisible water vapor from the spout soon condenses into water droplets that we see as steam. The same thing happens in nature on a vast

Below *During the ice age, 650,000 years ago, vast areas of Northern Europe were ice-bound, and may have looked much as Alaska does today.*

scale. Heat from the sun evaporates water vapor from the oceans and any wet surface. The water vapor in the air finally condenses in clouds to fall to earth as rain or snow. Because rivers soon return the rain and melted snow to the sea, the sea level changes very little over the centuries.

Ice age sea levels

During ice ages however, much of the snow that fell on northern lands remained as huge ice sheets instead of returning to the sea. This made sea levels lower than at present. We have found out how much lower in two ways.

The first way is to work out how much extra ice there was on land during the maximum of the ice ages. Using the known areas and average thicknesses of the glacial ice sheets, scientists have worked out the volume of the extra ice. This is found to be equal to a water layer 155 meters thick (508 feet) over all the oceans of the world. Therefore, when the ice sheets were at their largest, the sea level must have been around 155 meters (508 feet) lower than it is today.

Another way to work out how low the sea levels have been in the past is to explore the seabed to find the present depth of former beaches. It is not easy to find them by looking for them in a submarine. We can be more certain by finding remains of sea shells and plants that only grow along the seashore between high and low tide. There are many types of sea shells and many corals that grow here and are not likely to have been moved by the waves as the sea level rose. Scientists can measure the age of this material through **carbon-dating** samples of coral drilled from coral reefs, and shells dredged up from the sea floor. This tells us how much lower the sea level was at the time the corals or shells were alive. As this has been done in many places around the world, we now know that around 17,000 years ago, the average sea level was around 120 to 130 meters (395 to 425 feet) lower than at the present time. We also know from the age of these old beaches that some 15,000 years ago, the sea level rose rapidly from around 120 meters lower (395 feet) than present-day, until by 8,000 years ago, it lay within only 10 meters (33 feet) or so from its present level.

Naturally, 15,000 years ago, the basic

geography of many lands was quite different. Many shallow seas, such as the North Sea between Great Britain and Europe, became dry land, although this land was partly covered by an ice sheet. The Bering Straits between Asia and Alaska also became dry, which meant men and animals could cross from one continent to another. The Bass Strait between Tasmania and Australia, and many of the shallow tropical seas off Asia became dry land, so again men and animals could migrate to different lands.

Since the climate was colder at that time, the types of plants and animals in different lands also varied. Over tens of thousands of years there have been quite large changes in the life over the Earth. Particular plants and animals that now thrive in middle latitudes were found closer to the equator during the ice age. In the interglacial periods these plants and animals migrated away from the equator again.

Left Beneath the surface of the polar ice lie perfectly preserved traces of chemicals, dust, and air bubbles, perhaps thousands of years old.

Preserved in the ice

Because the polar ice sheets are very cold and contain little else but pure ice— much purer ice than you have in your refrigerator—the ice sheets provide frozen storage that preserves anything that falls on to the ice or becomes trapped in it. Air trapped in the polar ice up to 50,000 years ago is very similar to the present day air, and contains almost the same amount of oxygen and nitrogen. However, very careful work is needed to detect the very, very, small amounts of dust and chemicals in the ice.

By studying the ice cores from bore-holes, scientists have found that these impurities in polar ice have come from several places. There is quartz and similar dust that was picked up from other continents by the wind, and deposited on the ice sheets. Nearly ten times as much of this dust fell on the central part of the Antarctic ice sheet during the last ice age as is falling there at present. This tells us that it must have been quite windy at that time.

Salt spray from the oceans is also

found in the ice sheet, but the farther inland and higher one goes, the less sea salt is present. Dust and acid from volcanoes is also found in ice cores. While most of the volcanic dust falls out of the air during a few weeks after an eruption, the huge cloud of acid gas that is sent high into the air from the biggest eruptions encircles the globe and often takes a year or two to fall out.

The methods of studying ice cores are now so sensitive that we can also look for chemicals and dust pollution from

Pollution from distant countries may be carried by high clouds.

factories in countries far away from the ice sheets themselves. Radioactivity from the first nuclear bomb tests in 1955 has been measured in minute amounts all over the Greenland and Antarctic ice sheets. Lead from car exhaust fumes is found in increasing amounts in the Greenland ice formed from the snow that has fallen since 1950. It was at that time that leaded gas for cars became

Scientists taking ice cores from an ice shelf off Alaska.

much more common. However, this lead has not been found in the Antarctic ice sheet, as fewer people live in the southern hemisphere continents around Antarctica, and there are not so many cars south of the equator.

Debris from space

The largest objects found on the Antarctic ice sheets are **meteorites.** They have been collected elsewhere on earth for more than two hundred years, but up to 1970 only around 5,000 had been found. Since then, more meteorites have been found on the Antarctic ice sheet than all the meteorites found previously.

Most Antarctic meteorites have been found in two places. These are on areas of bare ice where the inland ice has been moving slowly against coastal mountain ranges for many thousands of years. Instead of glaciers flowing down to the sea, the ice in these areas is swept clear of snow by dry winds, which also evaporate ice from the surface. This exposes any small rocks carried to the surface, in the same way as the rocks of moraines were exposed below the snow line on mountain glaciers. However in these special areas in the Antarctic, there have been no mountains above the ice sheet to drop rocks onto the snow. The only rocks to fall on the snow are the very few that fall from space.

Among these rocks is one that looks like a special type of rock found only on the moon. It was probably a lump thrown off the moon when an even larger meteor, weighing many thousands of tons, hit the surface of the moon.

The Antarctic and Greenland ice sheets contain and cover many other secrets of nature which, like the meteorites, have still to be discovered. It is a difficult task to uncover these secrets, but the early explorers like Scott, Amundsen, Byrd, Wilkes, Mawson and many others were not daunted by the cold and winds of the polar regions. Their explorations set an example that has been followed by many others since, so that we are gradually learning more about the history of the earth, its glaciers and its ice sheets.

Right *New discoveries about the Earth's history are still to be made in the frozen wastes.*

Facts and figures

Scientists estimate that 6,020,000 square miles (15,600,000 square kilometers) or approximately 10.4 percent of the Earth's land surface, is permanently covered by glaciers.

The continent of Antarctica is as large as the United States of America and Europe put together, and twice the size of Australia. The coastline measures 14,000 miles (22,400 kilometers).

Antarctica contains 90 percent of the world's ice.

The ice covering Antarctica is so thick that only 2.5 percent of the land is visible.

The world's greatest thickness of ice was measured by radio echo soundings over Antarctica, recording a depth of 2.97 miles (4.7 kilometers).

Radio echo soundings have also revealed that there are many lakes under the Antarctic ice. The largest, near the center of the continent, measure approximately 3,088 square miles (8,000 square kilometers).

The Ross ice shelf in Antarctica covers an area larger than France. Seen from the sea, its edge is often a vertical ice cliff, towering up to 30 meters (nearly 100 feet) above sea level. Another 170 meters (560 feet) of ice may be hidden under the surface of the sea.

The longest known glacier, the Lambert Glacier, is in Antarctica. It is up to 40 miles (64 kilometers) wide, and including the upper section (known as the Mellor Glacier), is at least 250 miles (402 kilometers) long. Measured together with a further extension, known as the Fisher Glacier, the Lambert forms a continuous flow of ice about 320 miles (514 kilometers) long.

Under special circumstances, certain glaciers can "surge," advancing for short periods at a rate of several meters (or yards) per hour. When this happens, the forward movement of the glacier can be seen, and also heard, as the ground underfoot vibrates violently.

In 1982, the flow of the Quarayaq Glacier in Greenland was recorded to be 20 to 24 meters (65 to 80 feet) per day.

On a temperate glacier, where the annual snowfall may be as much as 5 to 10 meters (16 to 33 feet), it can take less than five years for fresh snow to be converted to glacier ice. On the much colder surfaces of the Greenland and Antarctic ice sheets, the same process can take around 3,000 years.

The tallest reported iceberg was seen off western Greenland in 1958. It was 167 meters (550 feet) high.

Icebergs can travel great distances if they drift into a strong ocean current. Off the west coast of Greenland, icebergs are carried south by the cold Labrador current, into the north-west Atlantic shipping lanes, where their movements must be carefully tracked if they are not to endanger the ships. These icebergs may cover as much as 12 miles (20 kilometers) a day and travel as far as 1,500 miles (2,400 kilometers) before melting away in warmer seas.

A floating island of thick Arctic ice was discovered in 1946. It measured 61 meters (200 feet) thick, and covered an area of 140 square miles (360 square kilometers), and was tracked for 17 years.

During the so-called "Little Ice Age" of the sixteenth to nineteenth century, the world's climate was much colder than it is today. Farmers in the Alps and the Scandinavian mountains were forced to leave the land as the advancing ice engulfed their grazing land, and even their farmhouses.

Glossary

Bedrock   The solid rock under a glacier or ice sheet.

Carbon-dating   A method of estimating the age of organic remains (such as wood etc.) by measuring the amount of radioactive carbon in the material.

Cirques   Steep-sided basin-shaped depressions on a mountainside, carved out by a glacier.

Crevasse   A large, deep crack in the surface of a glacier, caused by the ice flowing over a large bump in the bedrock.

Drumlins   Smooth rounded mounds of glacial till deposited under a glacier.

End moraine   The type of moraine which is deposited across the bottom end of a glacier.

Firn   A mass of large ice crystals in a glacier. This is the stage between loose snow turning into solid ice.

Fjords   Deep U-shaped coastal valleys, originally eroded by a glacier but now filled by the sea. Some fjords still contain a glacier.

Frost-shattering   Water trapped in cracks in the mountain rock expands as it freezes, shattering and breaking off pieces of rock.

Glacial till   Large amounts of soil, clay and rock that have been moved (sometimes over great distances) by an advancing ice sheet or glacier.

Glacier   A huge mass of ice, moving extremely slowly down a mountainside.

Ice Ages   Periods in the Earth's history when ice has covered a large part of its surface.

Icecap   A dome-shaped glacier usually covering a highland area. Icecaps are much smaller than ice sheets.

Ice sheet   A vast, thick layer of ice covering a large area of the Earth's surface, such as the Antarctic Ice Sheet, or the Greenland Ice Sheet.

Ice shelf   A thick floating ice sheet attached to a coast.

Ice streams   Fast-moving glaciers in an ice sheet which flow more rapidly than the surrounding ice sheet.

Interglacial periods The relatively warmer periods between each ice age, when the ice sheets retreat.

Medial moraine A central moraine formed where two glaciers join to form a larger glacier.

Meltwater Water produced by melting glacier ice, firn and surface snow. Meltwater flows down the bed of the glacier and emerges from the end as a stream often colored grey-green by the rock flour it contains.

Meteorites A rock-like object that has fallen from space onto the Earth.

Moraines As rocks are eroded by frost-shattering from the mountainside, they fall onto the surface of a glacier and are carried along on, or within the ice. Such rocks often form dark, parallel bands on the surface of the glacier and are called moraines.

Piedmont glaciers The parts of glaciers fed by mountain glaciers that have spread out over broad lowlands.

Rock flour The very finely ground rock particles which flow out from beneath a glacier in the meltwater.

Scree The steep slopes of frost-shattered rocks which pile up at the foot of a mountain.

Snow line A line on a temperate glacier, above which lies snow, and below which lies ice.

Tabular icebergs Large, flat, floating islands of ice, created when a large section of an ice shelf breaks away.

Temperate glaciers Those glaciers found in the more temperate climates of the world. The whole mass of ice in a temperate glacier warms to melting point in summer.

Further reading

Iceberg Alley by Madelyn K. Anderson (Messner 1976)

Glaciers: Nature's Frozen Rivers by Hershell H. Dixon and Joan L. Dixon (Dodd, Mead 1980)

The Antarctic by Pat Hargreaves (Silver Burdett 1981)

The Arctic by Pat Hargreaves (Silver Burdett 1981)

Icebergs and Glaciers by Patricia Lauber (Garrard 1961)

Beyond the Arctic Circle by George Laycock (Scholastic 1978)

The Arctic by D. Liversidge (Franklin Watts 1971)

Discovery in the Antarctic by S. McCullagh and L. Myers (Longman 1982)

Arctic Lands edited by Henry Pluckrose (Franklin Watts 1982)

Antarctica: The Great White Continent by Miriam Schlein (Hastings House 1980)

Index

Picture acknowledgments

The illustrations in this book were provided by: John Cleare/Mountain Camera 9, 20, 21; from Bruce Coleman—Gene Ahrens 11, Jen and Des Bartlett 22, 29, 34 Francisco Erize 25, 26, M. Freeman 12, Keith Gunnar *cover*, M. P. Kahl 27, Stephen J. Kraseman 17, Jarsolav Poncar 29, W. E. Ruth 18, WWF/G. W. Frame 41; Derek Fordham—Arctic Camera 10, 15, 24, 31, 33, 38, 40, 43; Photo Research International 4, 30; Alastair Stevenson/Mountain Camera 36; Zefa 7. The diagrams on pages 12, 14, 16, 28, and 32 were provided by Bill Donohoe.